CORROSION TECHNICAL SERIES
**NACE INTERNATIONAL TECHNOLOGY IN ACTION**

# Robotics

Volume I

# Other NACE Books of Interest

**Corrosion and Corrosivity Sensors**
Edited by V.S. Agarwala and G.K. Brown
Product Number: 37384

**Pipeline Coatings**
By Frank Cheng and Richard Norsworthy
Product Number: 37616

**Fusion-Bonded Epoxy: A Foundation for Pipeline Corrosion Protection**
By J. Alan Kehr
Product Number: 37580

**Field Guide to Internal Corrosion Mitigation and Monitoring for Pipelines**
By Richard Eckert
Product Number: 37610

**An Introduction to Corrosion Management in Industry**
By Ali Morshed
Product Number: 37633

Visit store.nace.org

CORROSION TECHNICAL SERIES
**NACE INTERNATIONAL TECHNOLOGY IN ACTION**

# Robotics

Volume I

**NACE International**
**The Worldwide Corrosion Authority**

©2020 by NACE International
All Rights Reserved
Printed in the United States of America

ISBN: 978-1-57590-403-0

NACE International
The Worldwide Corrosion Authority
15835 Park Ten Place
Houston, TX 77084
nace.org

**On the cover:** *Rolls-Royce believes that unmanned and remote-controlled transportation systems will become a common feature of human life. In one of its scenarios, data collected by UAVs could potentially enable engineers and inspectors to monitor marine structures from a remote, land-based control center. Photo courtesy of Rolls-Royce.*

# Contents

# Preface

Unmanned and remotely operated vehicles have been around for a long time and have served varied and important purposes, with technical improvements in both electronics and materials contributing to their increased usefulness across many industries.

In terms of the corrosion industry, Sprint Robotics likens what it calls "the rapid acceleration of the development and deployment of new technologies for the inspection, maintenance, and cleaning robotics sector" as essentially being driven by the multiple needs of asset owners, including improved efficiency and ensured human and environmental safety. According to data offered by Sprint, the global industrial inspection and maintenance market is valued at US$450 billion, and while that is the broad market figure, it clearly reflects the enormous opportunity and potential for implementing robotic solutions in all the industries that corrosion touches; which some could argue *is* all industries.

As implied above, there are many openings and opportunities for the increased use of robotic solutions, but the primary reasons seem to be to limit human intervention for both safety and accuracy. But they are not cheap. Just considering the global in-pipe inspection robotics market—according to data from Research & Markets—its value was around US$605 million in 2017 and it is expected to reach over US$2,450 million by the end of 2026, growing at a CAGR of more than 16.5% between 2018 and 2026.

It certainly is logical that in-pipe inspection robots would represent early embracers of the market. That is not an easy task for the human assigned to do it, or for the company that may need to face considerable downtime to complete the task.

Some of the more sophisticated in-pipe solutions include a real-time video feed back to a control station wherein the inspector can obtain an immediate sense of the pipeline conditions. While the investment for such an inspection solution may not be cheap, in the long run the accuracy, efficiency, and safety considerations seem to make the investment a wise choice.

To be sure, not all robots being used in the corrosion industry are crawling through pipe, some are crawling on, or flying above, the ground, determining the health and integrity of remote or hard-to-reach assets, providing feedback and enabling real-time decisions. In Chapter 1, "Corrosion Detection Using Robotic Vehicles," the overall problem faced by researchers from the Auburn University RFID Lab was to steer a robotic vehicle near to one or more inspection points to detect corrosion.

The team shows that a combination of algorithms including genetic algorithms (GA) for solving the traveling-salesman problem (TSP) and the dynamic window approach (DWA) are suitable for steering a differential-drive mobile robot to perform semi-automated inspections. Their simulation presupposes a map layout such as might be found on a utility right-of-way, or in a fenced-in area such as a utility substation. Machine-generated paths are compared with human-generated paths. It is shown that the human tendency to draw paths that prefer "wall following" can sometimes lead to suboptimal results, thus justifying the use of an iterative planner such as a TSP solver. They expect that a field-deployable inspection robot will have enough onboard processing power to use inputs from a camera, an IMU, and GPS for navigation. With these sensors and the algorithms described in this chapter, they show it should be possible to perform routine, automated inspections, thus saving time and money in the long run.

Chapter 2, "Underground Radar Detects Infrastructure Damage," takes the reader below the surface and presents an underground radar technology system developed at Louisiana Tech University that helped a coastal city identify and document underground infrastructure damage that had gone undetected in the years since Hurricane Katrina made landfall in south Louisiana in August 2005.

Known as FutureScan, this in situ pipe technology is a pipe-penetrating scanning system based on ultra-wideband (UWB) pulsed radar. The UWB system was first developed by the U.S. Department of Defense for the military, but a team at Louisiana Tech has refined it for civilian use. FutureScan allows for the inspection of buried pipelines, tunnels, and culverts to detect fractures, quantify corrosion, and determine the presence of voids in the surrounding soil often caused by storm water leaks and flooding, according to the developers.

While robots have already been used to find cracks and breaks in underground pipes, the new system looks past the pipe and into the dirt, in search of empty spaces and potential sinkholes. It can help civil engineers find voids as small as 8-in (203.2-mm) deep.

The initial research on the technology was funded by US$3.2 million from CUES, Inc., and US$3 million from the U.S. National Institute of Standards and Technology. Additional funds were provided by the state of Louisiana and from a US$400,000 research grant by the National Science Foundation.

Managing corrosion integrity for tanks and associated pipelines is a challenging and costly endeavor. To inspect tank floors, the tank's hydrocarbons or chemicals are usually emptied from the tank causing the tank to be out-of-service for an extended period. In addition to the monetary costs there is also a significantly increased safety risk.

The associated piping for transferring the hydrocarbons and chemicals in and out of the storage tanks are an additional integrity risk. The delivery lines are not designed for traditional in-line inspection (ILI) tools, often causing expensive modifications prior to ILI inspection. Chapter 3, "Managing Tank Integrity," explains how the attributes of "unpiggable" tools are designed to overcome this challenge and presents a case study courtesy of Intero Integrity that is associated with the inspection of a tank farm delivery line in 2017.

Chapter 4, "Robotic Crawler Turns the Unpiggable into Piggable," presents another case study, this time from Diakont Advanced Technologies in which the company was asked to determine the integrity of a natural gas pipeline. Part of the pipeline had never been inspected because it was partially buried in a densely populated area. Low flow, its narrow 10-in internal diameter, and its characteristics (tight bends, plug valves, etc.) made the pipe unsuitable for traditional smart pigging.

The solution was new robotic crawler tooling that could traverse challenging pipeline geometries using a ruggedized multiple track system, allowing for navigation across horizontal surfaces. The tool can extend the tracks to the pipe wall for stabilization. This arrangement provides the traction that is necessary to hold the tool rigidly in place while inspecting difficult-to access pipeline applications (such as inclines and vertical sections), where conventional ILI tools may not be feasible. Being self-propelled and bidirectional, the crawler can also be deployed and retrieved from a single access point, which was another key feature in its selection for this inspection program.

Cathodic protection (CP) site surveys are used to assess the effectiveness of corrosion protection on buried and submerged steel structures. It measures voltage gradients set up in the electrolyte by defects in coatings, which cause electrical currents to flow through the electrolyte surrounding the structure.

All CP site surveys require manual labor, and there is a huge possibility of both instrument and human errors involved in such surveys. Among these errors almost 90% is contributed by human errors and these errors rise due to many factors such as physical stress, lack of proper knowledge, improper handling of the survey equipment, etc.

In carrying out a survey, the surveyor must walk the entire pipeline route testing at regular intervals with the probes in a position of one in front of the other. The metallic structure may extend from a few meters to thousands of kilometers in size. Due to this, the survey becomes tedious and a huge stress is applied to the surveyor. By considering all these factors a manipulator robot was designed to perform the above-mentioned tasks. Chapter 5, "Automation for More Accurate CP Site Surveys," presents the conceptual design and some of the major applications of the same.

The conceptual robot design was developed to conduct CP site surveys in difficult and dangerous on- and off-shore locations. The user can control the robot from a remote location using a wireless teach pendant, and live survey data was sent to the inspector. The robot described in this chapter can perform ultrasonic crack detection, close interval potential survey (CIPS), direct current voltage gradient (DCVG) survey and global positioning system (GPS) mapping on buried structures, laid on both off shore and on shore.

Lastly, Chapter 6, "UAVs Take on Mission for Marine Corrosion, Coating Inspection," takes the reader in the air, and presents a few new partnerships that formed with the aim to expand drone, or UAV (unmanned aerial vehicle), use for maritime asset management. The concept of using UAVs for inspections is of interest to the maritime industry, since the marine environment presents numerous spaces that require either significant human risk or significant financial cost to access. However, as with many new technologies, challenges come with commercialization, costs, and processes.

To address those questions, several R&D projects launched in recent years are aimed at facilitating a more widespread adoption of UAV use to inspect for problems such as corrosion. These projects involve partnerships between industry, academia, and UAV technology groups—all designed to develop new end-to-end processes to enable the frequent use of drones to perform inspections in maritime settings.

By replacing human inspections with a UAV, routine maintenance can be monitored remotely and in real time by office-based staff, with instant feedback available to the vessel or offshore structure's superintendent. In turn, this can reduce costs, increase efficiency, and significantly reduce risks to human workers during essential maintenance. And that is everything that can be asked for any robotic or unmanned system for *all* the industries that corrosion touches.

# Corrosion Detection Using Robotic Vehicles

*Navigation algorithms developed at the at the Auburn University RFID Lab are presented that can be used to guide autonomous vehicles along a right of way to inspect poles or towers for corrosion, or to inspect enclosed spaces such as power substations. For testing, a differential-drive mobile robot was used in a large indoor space.*

Thaddeus Roppel, Yibo Lyu, Jian Zhang, and Xue Xia
Auburn University

Navigation generally consists of path planning, driving (path following), and obstacle avoidance. For autonomous vehicles, several important problems to be solved by the navigation algorithm are cost minimization (determining the most efficient path), graceful obstacle avoidance (implementing smooth curves rather than sudden stops and turns), determination of position with reference to a known map, and loop closure (*e.g.*, to know when a utility pole has been completely circumnavigated). The work presented in this chapter assumes the use of a visual inspection system, *e.g.*, a visible light camera. This may be supplemented with infrared and/or acoustic measurements without changing the essential strategies described here.

Several recent studies have indicated that dogs can be trained to detect corrosion by sense of smell.[1] The successful deployment of detection dogs and other animals in various applications, such as detection of drugs and explosives, has encouraged many researchers to attempt the design of electronic equipment that can mimic the sensory capability of the animal nose, while providing a less costly detection system with the capability to record quantitative data for real-time display and off-line analysis. Such a system is sometimes referred to as an "electronic nose" or simply an "e-nose." Currently, much research remains to be done to improve the sensitivity of electronic sensors to match the animal nose, thus our research focuses on developing autonomous mobile platforms that can drive themselves on rugged terrain while performing inspections in difficult areas.

The robotic vehicles studied here are guided mainly by known site maps, and can avoid obstacles in the process. The vehicles can provide real-time geotagged imagery, or store images to be offloaded at the management center for later analysis.

Corrosion detection is a key factor in any effort to mitigate the costs associated with corrosion in the utility power industry. A unique feature of corrosion detection is the requirement to get sensors as close as possible to the affected structural members. There may soon be portable mass spectrometers developed that can be deployed to aid in detecting corrosion below the soil surface.[2] However, so far no engineered remote sensor has been developed with similar capability. Thus, detecting corrosion requires close proximity visual inspection, or the application of electromagnetic methods at close range or in actual contact with the affected members.

This chapter addresses the issue of navigating an autonomous vehicle through a series of waypoints that are predetermined inspection points. These might be, for example, along a utility right-of-way, or inside a substation. The problem to be solved amounts to the inverse of the well-known coverage problem in robotics. In the coverage problem, the objective is to have the robot cover the entire free working space.[3] For example, a robotic vacuum cleaner should clean the entire floor without missing any area. However, in the present application, we want to inspect only the areas where corrosion can be present, such as the tower legs or poles near the ground contact points. In contrast to the coverage problem, the need here is to traverse as little ground as possible besides that which is essential to moving between inspection points.

In previous work, these researchers developed algorithms for just this type of situation. One previous application was to scan RFID tags in a retail sales floor, to count inventory.[4] The objective was to get as close to the merchandise display areas as possible to maximize count accuracy, while minimizing overall distance traveled around the sales floor. The algorithms developed are, essentially, solutions of the well-known traveling salesman problem (TSP), in which a salesman strives to visit all cities on a list without repeating his path and then return to the city of origin.

# Equipment Description

For algorithm testing, a differential-drive mobile robot was used in a large indoor space at the Auburn University RFID Lab. The waypoints were established by an initial manual drive-through, so that the locations were well established on a two-dimensional grid map. For algorithm development, Matlab software was used. Besides route planning, it is necessary to have an obstacle-avoidance strategy, accomplished using the dynamic window approach.[5]

The autonomous inspection vehicle might be required to navigate at three length scales:

- Pole to pole, roughly 100 m
- Around poles or tower legs, or within a substation, roughly 1 m
- Within close proximity to inspection points, roughly 0.1 m

The latter positioning would most likely be accomplished with a movable arm or gripper on which the inspection instrument(s) are located. This chapter is concerned with the first two scales–getting the robot from one pole or tower to the next, and then driving around the pole, or around each leg of the tower, or within a substation. It is also understood that the longest length scale (pole to pole) might most practically be accomplished by loading the inspection robot onto a truck and driving it, so that the only autonomous activity would be the second scale: around the pole or tower base, or within a substation.

From the navigation perspective, the two scales considered here have different requirements. For the large scale (long distance) motion, the primary objective is to move as quickly and directly as possible to the next inspection point (*e.g.*, pole or tower), while avoiding obstacles. For the medium scale, the objective is to drive around the pole or tower, with the speed and proximity determined by the instruments being used to detect corrosion. For example, to capture high-resolution still images, it might be necessary for the robot to stop at multiple times while circum-navigating a pole or tower leg. If the robot is carrying calibrated instruments, it might also be important to program a particular acceleration and deceleration pattern to avoid excessive jarring of the equipment.

The two most common types of equipment-carrying autonomous vehicles are (a) a differential drive vehicle with all equipment on board a single chassis, and (b) a tractor-trailer configuration, in which the motive power is provided by a tractor vehicle, and the equipment is carried on board an unpowered trailer chassis. For this chapter, only the single-chassis, differential drive configuration will be considered. The vehicle may be wheeled or tracked. The algorithms presented here will work for either, since they do not depend on highly precise odometery.

---

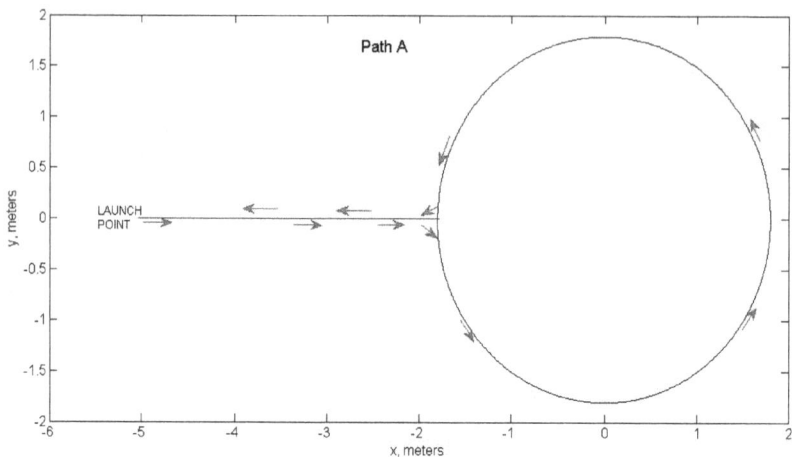

Figure 1.1: Path A – robot is released at the launch point (e.g., from a trailer) and programmed to make one loop around a pole. Pole diameter is assumed to be 0.6 m, robot track width is 1 m, and distance from inspection equipment on board the robot to the pole is 1 m. The path shown is the ground track of the center of the robot. Green arrows show motion away from the launch point. Red arrows show motion returning to the launch point. A key result is that the robot can identify when it has fully encircled the pole so it can return to the launch point.

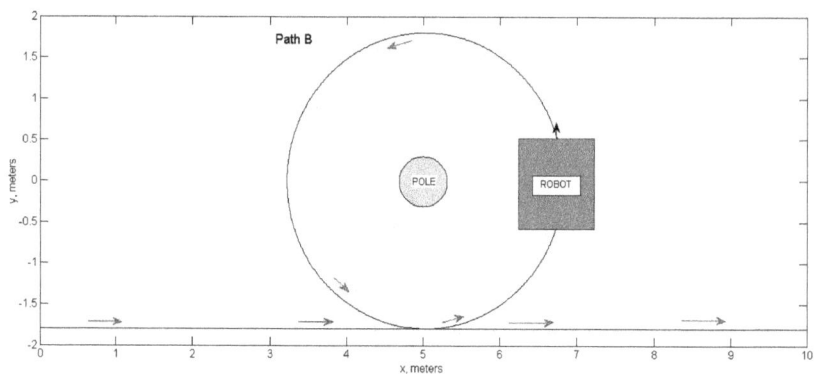

Figure 1.2: Path B – robot comes from previous pole, circumnavigates current pole, and continues to next pole.

A differential drive vehicle has two degree of freedom. It is able to turn on its axis, and move forward and backward. Fixed wheel velocities result in circular trajectories. For example, when the right wheel turns faster than the left wheel, the robot will follow a circular path in a counter-clockwise direction. One motion that is not possible for a differential drive vehicle is direct sideways motion. Standard automobiles suffer from this limitation as well. Like the automobile, a differential drive vehicle can ultimately assume any position and orientation from any given starting point, but the interim motions may be complex. (A familiar example of this for the standard automobile is parallel parking.)

Therefore, it is important to connect waypoints with realistic, achievable paths. It is also necessary to develop obstacle avoidance strategies that the robot can accomplish without undue strain on the drive system or on the sensing equipment which forms its payload. A fairly common strategy for obstacle avoidance for autonomous mobile robots is referred to as the previously referenced dynamic window algorithm (DWA).

DWA is designed to deal with the constraints imposed by limited velocities and accelerations for a short time interval. Trajectories are approximated by circular curvatures, resulting in a two-dimensional search space of translational and rotational velocities. Translational velocity is denoted by $v$, rotational velocity by Đ. In the search space only admissible velocities are considered, which ensures only safe trajectories are considered. A pair ($v$, Đ) is considered admissible if the robot is able to stop before it will collide with the closest obstacle. DWA chooses an optimal path in this search space for the robot to move from one goal to another. The optimal path will maximize a pre-determined objective function, such as a tradeoff between minimizing travel time and reducing rotational forces on attached equipment. The general form of the objective function is:

$$G(v,\omega) = \delta \left( \alpha \times heading(v,\omega) + \beta \times dist(v,\omega) + \gamma \times vel(v,\omega) \right)$$

Where,

- $heading(v,\omega)$ is a measure of progress toward the goal location
- $dist(v,\omega)$ is the distance to the closest obstacle on the trajectory
- $vel(v,\omega)$ is the forward velocity of the robot

and the function $\delta$ smooths the weighted sum of the three components and results in more side clearance from obstacles. The parameters $\alpha$, $\beta$, and $\gamma$ can be adjusted to prove the desired tradeoff.

To maximize sensor resolution, the robot needs to move as close as possible to the pole, so the local planner should choose a correspondingly suitable route in normal operation. It should also handle unexpected obstacles "gracefully," *i.e.*, without sudden, jerky motions.

Figure 1.3 shows an example in which there are five internal objects and a boundary fence to be inspected in a substation enclosure. The solution uses a multistep

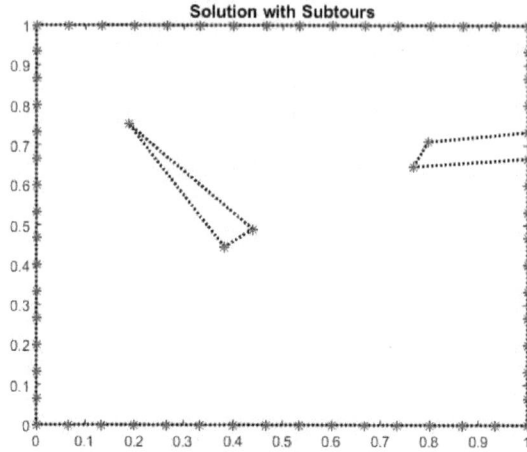

Solution with Subtours

process in which intermediate steps form so-called "subtours" – connected rings each containing a distinct subset of the set of all vertices.

Figure 1.3: Intermediate step in the TSP optimization of a set of vertices consisting of r~~........................~~ urs

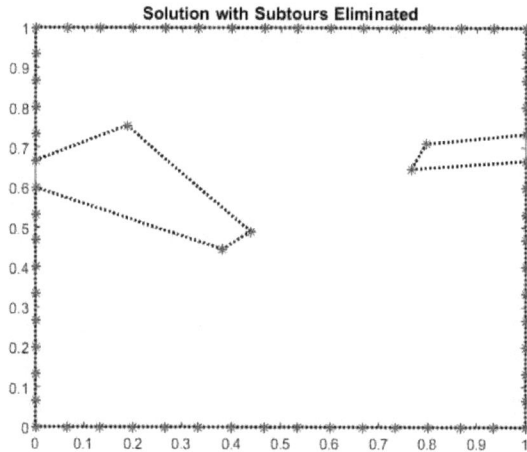

Solution with Subtours Eliminated

Figure 1.4: Solution of the TSP for this map with a single, optimal tour.
This map is the same as the one shown in Figure 1.3.

# Right-of-Way Inspection

As mentioned previously, this chapter focuses on differential-drive vehicles used to perform camera-aided visual inspection for corrosion. These vehicles need to perform two basic navigation functions: (1) Get to each pole on a list, and (2) drive in a circle around each pole. A third important capability is to avoid obstacles, such as hillocks, equipment, or workers that appear along the planned trajectory, whether temporarily or permanently. For this case, we do not yet have a suitable test facility, so the experiments described here are conducted in simulation. We do have several robotic vehicles currently being outfitted for deployment, and we hope to be able to report field data using those in the near future.

These experiments specifically show the ability of the robot to perform loop closure – that is, to identify when they have returned to their starting point on a circular trajectory. In an ideal case, this could be accomplished by employing some hypothetical "magic" GPS with infinite precision. Real GPS, however, has insufficient precision for the intended application, so additional information must be used if the robot is to make a smooth, complete circle around a pole. Most outdoor mobile robots use cameras and/or inertial measurement units (IMU) together with GPS for positioning and navigation. Sometimes wheel encoders are used as well, although they are of limited use on uneven and possible slippery terrain. Wheel encoders are even less useful for tracked vehicles, which can incur a substantial amount of slippage. The technique employed here in simulation mimics the use of a camera plus IMU by maintaining a fixed distance from the pole, and measuring both the range and heading back to the starting point of the circumnavigation.

As shown in Figure 1.1, Path A is based on deploying the robot from a trailer or truck bed, assigning it to inspect one or more poles, and returning. The strategy employed for this, for inspecting a single pole, is composed of five segments:

- Drive straight toward the center of the pole.
- Upon reaching the desired proximity to the pole, rotate in place 90 degrees.
- Begin driving a circular path by setting the right and left wheel speeds to the appropriate constant values
- Upon returning to the start point, rotate in place 90 degrees to face the trailer.
- Drive straight back to the trailer.

Each of these segments will contain error in real life. Slight differences in motor responses to computer commands result in trajectory error, which must be compensated sufficiently to ensure complete inspection, and returning to base "close enough."

To study the effects of random error, a variety of paths were generated with cer-

---

tain amounts of random error introduced in the actuation model. For real-world operation, this would be primarily due to motor speed variations and terrain effects. Figure 1.5 shows a collection of these paths generated at random, each with

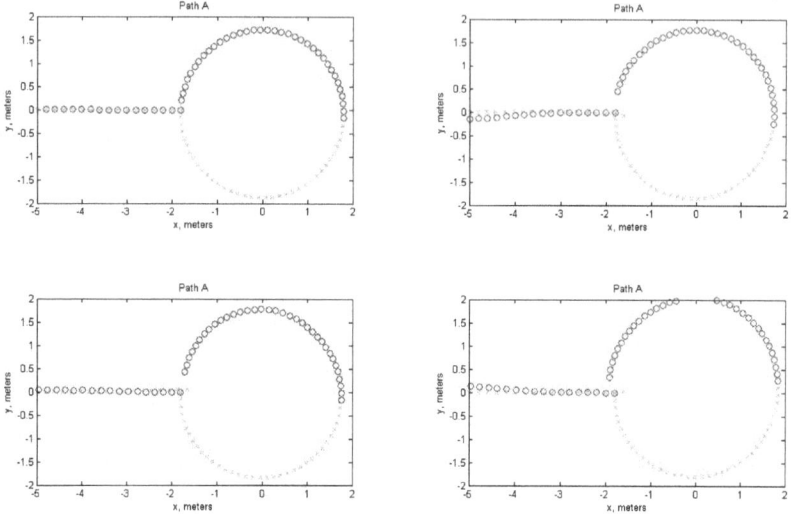

the same, relatively small, random error distribution in the wheel velocities.

Figure 1.5: Upper left: Ideal path traversed by the inspection robot. Green x's are the outbound part of the path. Red circles show the returning part of the path. With slight errors introduced into the actuation, paths such as those shown in the other panels are obtained. The primary effects are lack of precise loop closure, and non-straight approach and return paths. It can be seen that the path deviates from a perfect circle as well.

Increasing the amount of actuation error results in further deviation from the ideal path. However, in all but the most extreme cases, the robot is still able to complete a loop around the pole and return close enough to the launch point to be recovered with minimal effort. Figure 1.6 shows some typical results from intro-

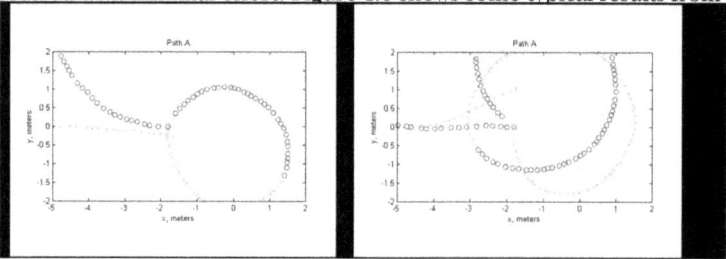

ducing a high level of actuation error.

Figure 1.6: Several representative cases showing extreme deviation
resulting from very large actuation error.

Path B simulation results are similar to those for Path A. For this path, the robot approaches a pole, either from a previous pole, or from the launch point, encircles the pole, and then continues toward the next pole, or to the recovery point. One additional factor of concern for Path B is that the heading deviation upon leaving a pole can have severe consequences if not corrected, since otherwise the robot may never find the next pole. The required heading correction can be achieved readily using an onboard IMU.

## Substation Inspection

Experiments were conducted in simulation using the Matlab Optimization Toolbox. A simulated map was created representing a substation, or in fact any enclosed space for which interior and perimeter inspection is desired. Inspection points were distributed at evenly spaced locations on the periphery (*e.g.*, fence posts) and at up to 10 locations in the interior. Paths were generated using solutions to the TSP.[7] For physical experiments, a differential-drive mobile robot was operated inside a laboratory environment containing desired inspection points, as well as obstacles to be avoided.

Figure 1.7 shows results from a solution of the TSP applied to a bounded rectan-

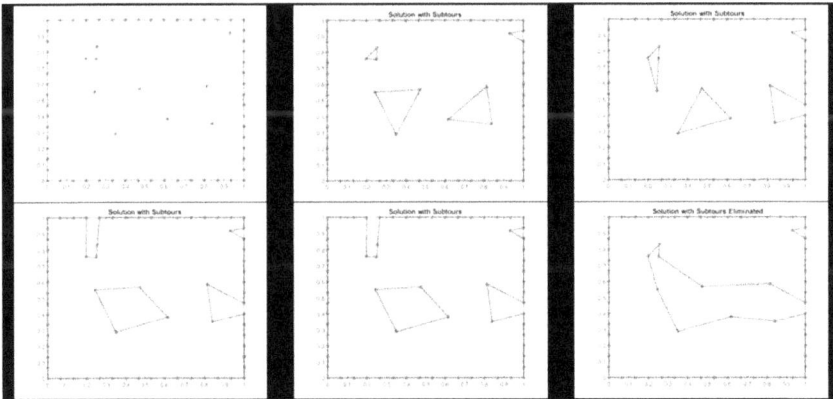

gular enclosure with both interior and perimeter points to be inspected.

Figure 1.7: TSP applied to a bounded rectangular enclosure with both interior and perimeter points to be inspected. The red stars are inspection points. These are connected by blue lines indicating proposed paths for the robot to follow. The upper left panel shows the original map.

The lower right panel shows the solution. Intermediate panels show subtours generated during the solution using a genetic algorithm approach.

Often, the machine-generated solution is not the one that a human would natu-

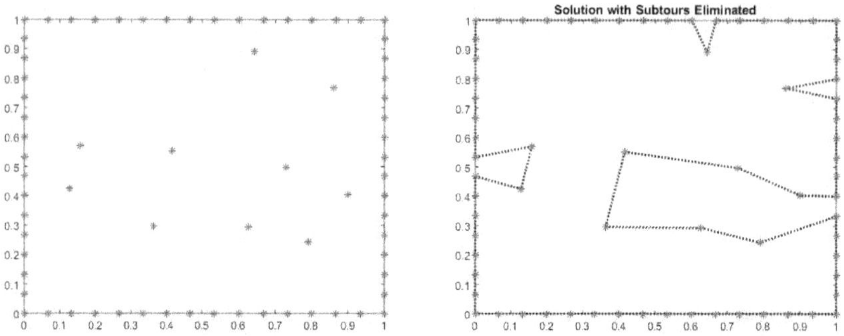

rally draw, but the computational process generally results in the minimum cost solution so long as total travel distance is the only criterion. To illustrate this, several individuals unfamiliar with the research were asked to draw what they felt were the best paths in a randomly generated map. Their results were compared with the machine-generated solution. Figure 1.8 shows the original map, and the machine-generated solution.

Figure 1.8: Original map (left panel) and machine-generated solution (right panel).
Red stars are inspection points. Blue lines indicate the proposed path of the robot.

For comparison, Figure 1.9 shows a collection of path drawings generated by five individuals. Although each path is different, there are commonalities. The prefer-

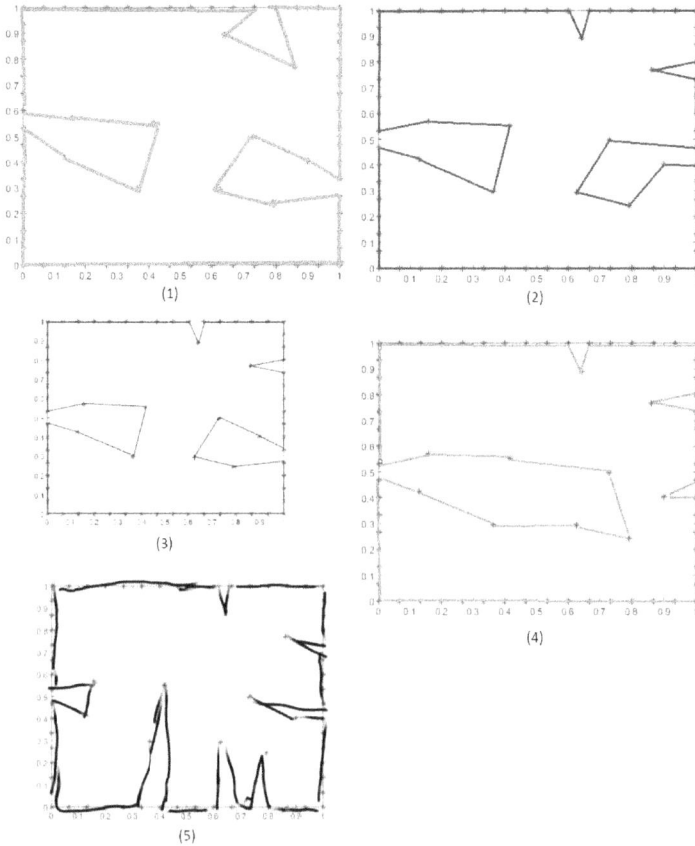

(1)

(2)

(3)

(4)

(5)

ence for human-generated paths seems to be to avoid large deviations from the current position on the perimeter in order to explore the interior. In contrast, the machine-generated solution is more willing to detour deeply into the interior if that results in the overall minimum path length. This is an expected and predictable consequence of the iterative solution process that is able to generate and rank quantitative intermediate results.

Figure 1.9: TSP path tours of the same map generated by five different individuals.

The overall problem addressed in this work was to steer a robotic vehicle near to one or more inspection points to detect corrosion. We found that a combination of algorithms including genetic algorithms for solving the TSP and the DWA are suitable for steering a differential drive mobile robot to perform semi-automated inspections.

Our simulation presupposes a map layout such as might be found on a utility right-of-way, or in a fenced-in area such as a utility substation. Machine-generated paths compared with human-generated paths showed that the human tendency to draw paths that prefer "wall following" can sometimes lead to suboptimal results, thus justifying the use of an iterative planner such as a TSP solver.

We expect that a field-deployable inspection robot will have sufficient onboard processing power to use inputs from a camera, an IMU, and GPS for navigation. With these sensors and the algorithms described here, it should be possible to perform routine, automated inspections, thus saving time and money in the long run.

*This chapter is based on and imagery is taken from NACE International Technical Paper C2017-9741, "Corrosion Detection using Robotic Vehicles in Challenging Environments," by Thaddeus Roppel, Jian Zhang, Yibo Lyu, and Xue Xia, Auburn University. To view the paper in its entirety, visit store.nace.org.*

# REFERENCES

1.  A. Schoon, R. Fjellanger, M. Kjeldsen, K. Goss, "Using Dogs to Detect Hidden Corrosion," *Applied Animal Behaviour Science* 2014, 153 (2014): pp. 43-52.
2.  M. Yang, T. Kim, H. Hwang, "Development of a Palm Portable Mass Spectrometer," *J. American Society for Mass Spectrometry* 19, 10 (2008): pp. 1442-1448.
3.  A. Zelinsky, "A mobile robot exploration algorithm," *IEEE Transactions on Robotics and Automation* 8, (1992), pp. 707–717.
4.  J. Zhang, Y. Lyu, T. Roppel, J. Patton, C. Senthilkumar, "Mobile robot for retail inventory using RFID," *2016 IEEE International Conference on Industrial Technology (ICIT)*, pp. 101-106.
5.  D. Fox, W. Burgard, S. Thrun "The Dynamic Window Approach to Collision Avoidance," *IEEE Robotics & Automation Magazine* 4, 1 (1997), pp.23–33.
6.  D. Applegate, R. Bixby, V. Chvátal, W. Cook, *The Traveling Salesman Problem*, Princeton University Press (2006).
7.  X. Xia, "Robot Navigation for RFID-based Inventory Counting," Master's Thesis, Auburn University, 2016.

# Underground Radar Detects Infrastructure Damage

*In situ pipe technology created at Louisiana Tech's Trenchless Technology Center incorporates simulation, electronics, robotics, signal processing, and three-dimensional renderings in a package that can be mounted on existing pipe-inspection robots.*

Ben DuBose
Materials Performance

A new underground radar technology system developed at Louisiana Tech University helped the coastal city of Slidell identify and document underground infrastructure damage that had gone undetected in the years since Hurricane Katrina made landfall in south Louisiana in August 2005.

Known as FutureScan, this in situ pipe technology is a pipe-penetrating scanning system based on ultra-wideband (UWB) pulsed radar. The UWB system was first developed by the U.S. Department of Defense for the military, but a team at Louisiana Tech has refined it for civilian use.

The system allows for the inspection of buried pipelines, tunnels, and culverts to detect fractures, quantify corrosion, and determine the presence of voids in the surrounding soil often caused by storm water leaks and flooding, according to the developers.

While robots have already been used to find cracks and breaks in underground pipes, the new system looks past the pipe and into the dirt, in search of empty spaces and potential sinkholes. It can help civil engineers find voids as small as 8-in (203.2-mm) deep.

## UWB Technology

Created at Louisiana Tech's Trenchless Technology Center, the technology incorporates simulation, electronics, robotics, signal processing, and three-dimensional (3-D) renderings, all in a package that can be mounted on existing pipe-inspection robots.

The radar device, housed in a casing the size of a smartphone box, is strapped to the top of a robot (Figure 2.1) and integrated into existing assessment software. As the wheeled robot travels through the pipeline, it gathers video footage, sends out signals, and processes the reflection, thus giving the operator a detailed report with a graphical display of pipe anomalies and the likelihood of voids.

Figure 2.1: The radar device is strapped to the top of a pipe-inspection robot and integrated into existing assessment software. Photo courtesy of Louisiana Tech University.

Arun Jaganathan, associate professor of civil engineering and construction engineering technology at Louisiana Tech, began developing this technology in early 2009 as the basis for his Ph.D. dissertation research. Partnering with fellow Louisiana Tech researcher Neven Simicevic and others, his vision was to develop it into a tool that municipal engineers can use for their routine pipeline condition assessments.

"Our UWB technology was based on recognizing the need within the trenchless industry for an advanced pipeline inspection tool that can quantify the structural integrity of buried municipal pipes like sewers and storm drains, and be able to see through the pipe wall," said Jaganathan.

"The radar system emits ultra-short electromagnetic pulses from inside of a sewer pipe and captures the signals back-scattered from the pipe to determine the condition of various layers hidden behind the wall, which we cannot directly see using visual tools such as a camera," he said. "The radar is integrated into a robot, which crawls through a pipe and relays the data back to the operator in real time."

Figure 2.2: While robots have already been used to find cracks and breaks in underground pipes, the new system looks past the pipe and into the dirt, in search of empty spaces and potential sinkholes.

Although other robots have used ground-penetrating radar, the antenna on those existing robots must touch the pipe; however, the antenna used on this system does not.

# Application Tests

Jay Newcomb, a Slidell city council member and Louisiana Tech alumnus, first learned of the technology through connections with his alma mater.

"We took a trip to [Louisiana] Tech in September of 2010 to check things out and make ourselves be known to any interested companies," said Newcomb. "At that time, Louisiana Tech's radar technology was still in the developmental stages. But the research team said that if the innovation proved useful in lab tests, Slidell would be used as a beta site in actual field studies."

Following successful testing and development, Jaganathan and other researchers came to Slidell in the summer of 2013 to pinpoint the spots in the city that would be most beneficial for using the UWB, and to test and investigate the underground infrastructure issues. As predicted by the group in their initial research, compromised infrastructure could be seen when using the UWB technology.

"While we were aware of the depth and breadth of the problems that plagued our underground utilities and we knew surrounding communities had experienced similar problems, I believe it wasn't until we made the trip to Ruston in 2010 and then saw the results of the UWB investigation that we actually realized we could have quantifiable evidence of the scope of that damage," said Newcomb.

According to Jaganathan, the technology is distinctive from other radars on the market.

"This technology is unique in its capability to generate high resolution images, which allow engineers to inspect a particular spot in detail," said Jaganathan. "Unlike many other radars, our system does have to be in contact with the pipe wall, and this provides capability for rapid inspection to finish scanning a long pipe in a timely manner."

As a result of the work of Jaganathan, Simicevic, the Louisiana Tech research team, and consultants with other engineering firms, Slidell was able to secure $75 million in funding from the U.S. Federal Emergency Management Agency to begin the underground utility restoration process.

Figure 2.3: According to Arun Jaganathan, shown in his lab, this technology is unique in its capability to generate high resolution images, which allow engineers to inspect a particular spot in detail. Photo courtesy of Louisiana Tech University.

"Our fiscal 2017 total budget for the City of Slidell is just under $43 million," said Newcomb. "We now have almost two whole budgets to spend on streets, drainage, and sewer, thanks to the collective efforts of many, beginning with the research conducted by Louisiana Tech."

After the successful test in Slidell, Louisiana Tech applied for and received a U.S. patent on the technology. Commercial developers with pipeline inspection company CUES, Inc. say they are continuing to refine the radar system to make it simpler and easier to use for clients.

## Template for Future Projects

Jaganathan believes the success of the project demonstrates the value of academic research.

"What started as an academic research ultimately led to the development of a practical tool that our municipal engineers can use on a daily basis for the betterment of our infrastructure and society, as a whole," he said.

In a state like Louisiana, which is often prone to hurricane damage based on its position on the U.S. Gulf Coast, the tool could prove quite useful in the years ahead. From a policy standpoint, Newcomb believes Louisiana would benefit from keeping a closer eye on technical research from local universities.

"I truly believe that far too many municipalities' first response to problems is 'Who do we hire to consult/fix this?' instead of asking 'I wonder if any of our public universities have researched this topic or have any prior experience dealing with a similar situation?'" said Newcomb.

The initial research on the technology was funded by $3.2 million from CUES, Inc., and $3 million from the U.S. National Institute of Standards and Technology. Additional funds were provided by the state of Louisiana and from a $400,000 research grant by the National Science Foundation.

*This chapter is based on the Materials Performance article, "Researchers Use Underground Radar to Detect Post-Storm Infrastructure Damage," by Ben Dubose. To read the article in its entirety, visit www.materialsperformance.com.*

# Case Study: Managing Tank Integrity

*Out-of-service tank inspection is a costly, inefficient way to manage tank floor integrity. Periodic out-of-service inspections are still needed but improvements in tank robotics now allows for in-service risk-based inspections, allowing tank operators to minimize safety risks and optimize rehabilitation planning and maintenance.*

Mark Slaughter
Intero Integrity

---

Recent developments in robotic technology have allowed operators to extend the service on above-ground storage tanks, in adherence with regulations, before taking the tank out of service. This extends the uptime of tanks and reduces costs and safety risk.

When Philips Petroleum made a decision to inspect four pipelines at a tank farm in Torrance, CA, they determined the pipelines were considered unpiggable. With no pig traps on the line, no space to install temporary traps, and no previous inspection data, there was a risk of having significant problems during the inspection. For this reason, they hired Intero Integrity to inspect the lines utilizing their Piglet fleet of tools.

As the pipelines were located in a densely populated urban area, Phillips needed a way to inspect the lines within budget. The company's tools were bi-directional, saving significant costs by allowing for a single entry/exit of the inspection tool. The inspection tools are ultrasonic tools, the most accurate technology for measuring pipeline wall loss.

Prior to mobilizing for the inspection, it was crucial to know the requirements for project. For this reason, a site visit was made and the inspection company provided temporary valves, flow meters, pressure gauges, and a temporary pig trap. Two portable pumps and a VAC truck were provided for pumping the tool in both directions. The tool was launched from the temporary pig trap and was pumped with diesel. Prior to running the smart pig three cleaning tools were run through the pipelines.

The inspections were considered a success by Philips. Each bi-directional run gave the company not one, but two data sets to evaluate the condition of the pipelines. This is especially useful when there is echo loss due to residual dirt and debris. Having two data sets adds enormous value in the situations.

# Out-of-Service vs. In-Service Inspection

Improved electronics over the last several years have enhanced tank floor inspections. Improved ultrasonic (UT) inspection methods address storage tank integrity monitoring and assessment without removing tanks from service. By acquiring large amounts of high-density UT data and evaluating them with readily available analysis tools, inspectors can provide tank owners and regulators with insight into the integrity of above-ground storage tank (AST) floors not otherwise available.

In-service robotic inspection of tank floors is now readily accepted by the industry. The cost of taking a tank out of service can be high, including downtime, safety risk, and associated labor and rental costs for transferring of the tank product. Also, in-service tank inspection can assist operators in planning future outage time, material, and manpower resources. By determining tank integrity, through validation of the EVA (Extreme Value Analysis) statistical lifetime assessment, in-service robotic inspection is taking the technological lead in managing tank integrity.

In general, it will not be possible for a remotely operated vehicle (ROV) to reach/ cover 100% of a tank bottom during an inspection. Typical areas that won't be covered are shown in Figure 3.1. In principle, areas behind a fixed structure, seen from the manhole through which the ROV has been deployed, cannot be inspected. These areas are not accessible for the ROV due to size and safety operations procedures.

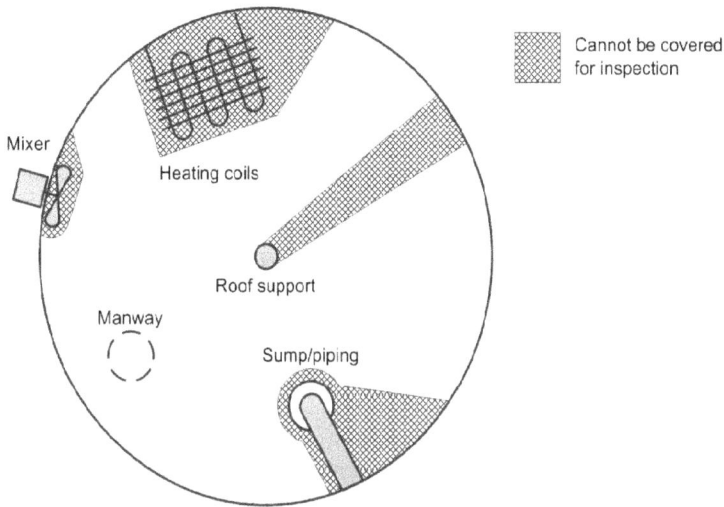

Figure 3.1: Example of typical tank inspection limitations.

# Deployment and Inspection

The deployment of an ROV is usually carried out through a manhole on the roof of the tank. A specially designed tripod is used to lower the ROV into the tank and special guidance is required for the umbilical. The length of the control cables is often limited, so the control area needs to be as close to the tank as possible. Special safety precautions are applicable to work on the tank roof (fall protection). When the ROV is lowered to the bottom of the tank, it is maneuvered around in the tank following a pre-determined inspection route. The purpose of the inspection plan is to ensure enough coverage of the tank bottom and minimize the risk of the ROV becoming entrapped

In general, the following list of tank conditions must be confirmed to perform a quality and accurate inspection.

- Ultrasonic properties of the existing floor coating must be known or determined. Air gaps/inclusions could influence the measurements.
- The manhole must be free of any obstruction and have a minimum diameter of 20 in (508 mm) for the ROV.
- Product temperature should not be higher than 110°F.
- Product in the tank should be filled to a pre-determined level (based on tank characteristics).
- The floor must be free of any obstruction such as anodes, heating coils, etc. Any obstructions should be mapped out in the inspection route plan to determine a safe working area and driving procedure.

---

# Extreme Value Analysis

Extreme value analysis (EVA) is a statistical tool used to estimate the likelihood of the occurrence of extreme values based on observed/measured data; this is an accepted practice for assessing the minimum remaining metal thickness of the tank bottom. The number of measurements taken for a statistical sampling will depend on the size of the tank and the degree of expected soil-side corrosion. Typically, 0.2% to 10% of the bottom should be scanned for a representative evaluation used for inspection data know as partial coverage inspection (PCI), where access, cost, or other limitations result in an incomplete dataset. In PCI, EVA can be used to estimate the largest defect that can be expected.

Due to time, internal tank obstructions, and cost, UT inspection of 100% of the tank floor is not feasible. Most inspections measure floor thickness over a well-distributed percentage of the floor and then estimate minimum floor thickness using EVA. Some tank owners also use EVA statistics when they evaluate out-of-service floor UT data following MFL scans.

Presuming the inspection plan meets requirements for application of EVA, data needs to be collected from only a small percentage of the tank floor. This test method has been validated and that there is no quantifiable difference in determining remaining life for the tank floor inspection as long as the data collected is distributed among all the navigable tank floor plates. Results of field tests and more than 20 independently monitored validation studies demonstrate that a small population sample of the tank floor can provide satisfactory results.

# Managing Explosion Risk

When will an explosion occur? The most critical time is when the ROV is being deployed through the vapor zone. The most common types of reactions are between flammable gases, vapors, or dust with oxygen contained in the surrounding air. This is a serious discussion for people working in environments such as in and around above ground storage tanks. Careful consideration must be made to the specifications of the tank and product as well as the design considerations of the tank inspection equipment to eliminate all risks of ignition.

There are three basic requirements must be met for an explosion to occur in atmosphere air:

1. **Flammable substance**—needs to present in sufficient quantity to produce ignitable or explosive mixture.
2. **Oxidizer**—must be present in sufficient quantity in combination with the flammable substance; the most common is $O_2$.
3. **Source of Ignition**—a spark or high heat must be present.

---

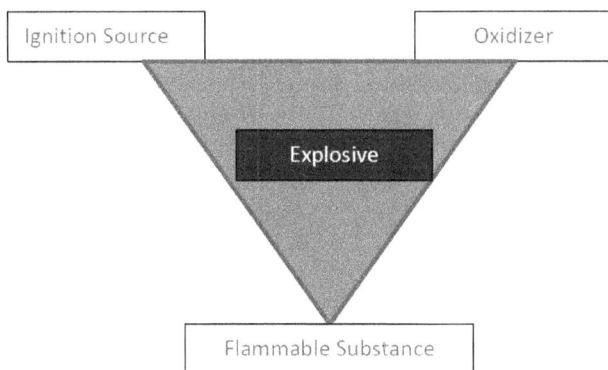

Figure 3.2: The presence of these three elements make up the sides of an ignition triangle.

| Flammable Substance | Examples | Description |
|---|---|---|
| Flammable Gas | Hydrogen, etc. | Often compounds of hydrogen and carbon that require very little to react with atmospheric oxygen |
| Flammable Liquids/Vapors | Hydrocarbons such as ether, acetone, lighter fluids, etc. | Even at room temperature, sufficient quantities of the hydrocarbons can evaporate to form a potentially explosive atmosphere at their surface. |
| Flammable Solids | Dust, fibers, and flyings | The cumulative nature of dust hazard is the most significant difference between gas/vapor and a dust hazard.<br>A dust cloud will settle on nearby surfaces if it is not ignited.<br>Unless removed, layers of dust can build up and will serve as fuel for subsequent ignition.<br>The typical dust explosion starts with the ignition of a small dust cloud resulting in relatively small damages.<br>Pressure waves of the small initial explosion are the most damaging part of the dust explosions.<br>These pressure waves release dust layers from surrounding vertical or horizontal surfaces to produce a larger cloud with is ignited by the burning particles of the initial cloud |
| Oxidizer | Air at normal atmospheric conditions | When the amount of available atmospheric oxygen is more or less in equilibrium with the quantity of flammable material, the effect of an explosion – both temperature and pressure – is most violent |
| Ignition source | The amount of energy required to cause ignition is dependent upon these factors: | 1. The concentration of the hazardous substance within its specific flammability limits.<br>2. The explosive characteristics of the particular hazardous substance.<br>3. The volume of the location in which the hazardous substance is present. |
| Ignition Sources (Industrial Electrical Equipment) | Hot Surfaces: Electrical Sparks Friction and Impact Sparks | 1. Surfaces heated by coils, resistors, lamps, brakes or hot bearings. Hot surface ignition can occur at the Auto-Ignition Temperature (AIT) or spontaneous ignition temperature at which a hazardous substance will spontaneously ignite without further energy.<br>2. Occurs when circuits are broken or static discharge takes place In low voltage circuits. Arcs are often created through the making and breaking of electrical contacts.<br>3. When casings or enclosures are stuck. |

There are three principles to follow that ensure electrical equipment does not become a source of ignition. The basic point is to ensure that parts to which a potentially explosive atmosphere has free access does not become hot enough to ignite an explosive mixture.

The design of robotic tank inspection systems depends on the individual characteristics of tank and its product. For example, a water storage tank with no residual hydrocarbons or chemicals will require less design safety features than with a gasoline tank.

| No. | Principles | Protection Method |
|-----|-----------|-------------------|
| 1 | **Confine explosion**<br>Explosive mixtures can penetrate the electrical equipment and be ignited. Measures are taken to ensure that the explosion cannot spread to the surrounding atmosphere. | Explosion-proof enclosure<br>Dust ignition-proof enclosure<br>Conduit and cable seals |
| 2 | **Isolate the hazard**<br>The equipment is provided with an enclosure that prevents the ingress of a potentially explosive mixture and/or contact with sources of ignition arising from the functioning of the equipment. | Pressurization and purging<br>Oil immersion<br>Hermetic sealing<br>Encapsultation (potting)<br>Restricted breathing |
| 3 | **Limit the energy**<br>Potentially explosive mixtures can penetrate the enclosure but must not be ignited. Sparks and raised temperatures must only occur within certain limits. | Intrinsic safety<br>Pneumatics<br>Fiber optics |

Figure 3.3: Shown is the ES350 ROV used by Intero Integrity.

In July 2017 Intero Integrity performed its first in-service inspection in the U.S. with the E350 robot. This was a feasibility study conducted on an aboveground 3.3 million-gallon fuel storage tank, and was the culmination of two years of planning by the Defense Logistics Agency, Fleet Logistics Center (FLC) Jacksonville. The robot was able to detect elevation changes on the tank floor, thereby identifying ground erosion. Moreover, the corrosion anomalies on the tank floor that were detected and sized were within 0.1% accuracy from the subsequent manual UT sizing.

Figure 3.4: Intero Integrity's E350 robot and floaters being lifted onto a tank roof.

Intero was involved in a second U.S. project using its remotely operated E350 robot. That project was commissioned by the Tennessee Valley Authority (TVA), located in Gallatin, Tennessee. The tank's dimension were 138' in diameter x 48' tall and it stored No. 2 diesel fuel. TVA had a regulatory requirement to inspect this tank. The cost to perform an out-of-service inspection of the tank floor would have been significant. The use of the ROV allowed Intero Integrity not only to charge significantly less for the inspection, but it was able to complete the entire project in just three days.

*This chapter is based on and imagery is taken from NACE International Technical Paper C2019-13340, "Overcoming Corrosion Integrity Management Challenges for Storage Tanks and Associated Pipelines," by Mark Slaughter, Intero Integrity. To view the paper in its entirety, visit store.nace.org.*

# Case Study: Robotic Crawler Turns the Unpiggable into Piggable

*Diakont Advanced Technologies was commissioned to assess the integrity of a natural gas pipeline that was partially buried under an urban area on a major North American pipeline. The company used a reduced size robotic crawler to successfully navigate a 10-in pipe. The size of this pipe has previously been a limitation, making it "unpiggable" using other ILI methods.*

Jonny Minder
Diakont Advanced Technologies

Engineering firm Diakont Advanced Technologies used a robotic crawler to assess the integrity of a natural gas pipeline partially buried under an urban area for a major North American pipeline that had previously been considered "unpiggable" using other in-line inspection (ILI) methods.

Designated as a high consequence area (HCA) due to being located in a densely populated urban area, this section of pipeline had never been inspected. Its low flow, narrow 10-in internal diameter (ID), and pipeline characteristics (e.g., tight bends, plug valves) made it unsuitable for traditional smart-pigging.

However, U.S. Pipeline and Hazardous Materials Safety Administration (PHMSA) regulations require specific integrity management programs in HCAs. The piping's inspection challenges could have forced the pipeline operator to replace the entire quarter-mile length of pipe if they could not inspect the line on schedule. The technology gap between inspection requirements and available tooling forced the industry work with pipeline service vendors to develop a solution.

# A Reduced-Size, Self-Propelled Solution

The HCA consists of 1325 ft of partially buried line located near a library, a school, and private homes. Again, its low flow, pipeline geometry, and a lack of sufficient access points prevented the use and retrieval of a free-flowing smart pig. For these reasons, an ILI solution was deemed necessary, but most existing self-propelled ILI solutions readily available on the market were too large to navigate the piping's narrow ID. The operator determined that a smaller form-factor inspection crawler solution was needed to navigate the piping's challenging geometry (including vertical sections and mitered bends) as well as to allow entry and retrieval from a single access point without the large footprint of a pig launcher/receiver.

Faced with these inspection challenges, the pipeline operator reached out to Diakont for a solution. The operator selected Diakont's Sprinter robotic crawler tool for the inspection because of its ability to negotiate unpiggable pipeline geometries with IDs as small as 8 in. The Sprinter accommodates small IDs by arranging the robotic crawler track modules, control electronics, and NDE sensor suite modules into a linear assembly. The robotic modules attach together via rugged flexible connections, contributing to the Sprinter's increased maneuverability. The Sprinter inspection system is mobilized on-site in separate sections and assembled near the pipe access point to reduce the space required for the inspection crew and the impact to the pipeline operator.

Sprinter is able to traverse challenging pipeline geometries using a ruggedized multiple-track system for navigation across horizontal surfaces, and the tool can extend the tracks to the pipe wall for stabilization. This arrangement provides the necessary traction to hold the tool rigidly in place while inspecting difficult-to-access pipeline applications where conventional ILI tools may not be feasible, including inclines and vertical sections. The Sprinter system moves at a deliberate pace to provide accurate mapping of anomaly locations within the pipeline. Self-propelled and bidirectional, the Sprinter can also be deployed and retrieved from a single access point, which was another key feature in its selection for this inspection.

Diakont's Sprinter tool is equipped with a non-destructive examination sensor suite to provide comprehensive assessments of pipe wall conditions:

- Diakont's patented ultrasonic testing (UT) electromagnetic acoustic transducer (EMAT) module measures pipeline wall thickness.
- High-definition cameras allow technicians to gather photographs and video of the appearance of the pipeline's interior.
- All sensor data feeds back to the inspection technicians in real time via an umbilical cable to permit same-day assessment results.

Unlike conventional UT instruments that employ contact methods, the EMAT instrument does not require any liquid couplant between the transducers and pipe wall to transmit ultrasonic signals.

---

This noncontact method is ideal for gas pipeline inspections, where liquid within lines is unacceptable. Figure 4.1 illustrates the principle of EMAT inspection, where an EMAT direct-beam (0°) transducer excites and receives shear waves.

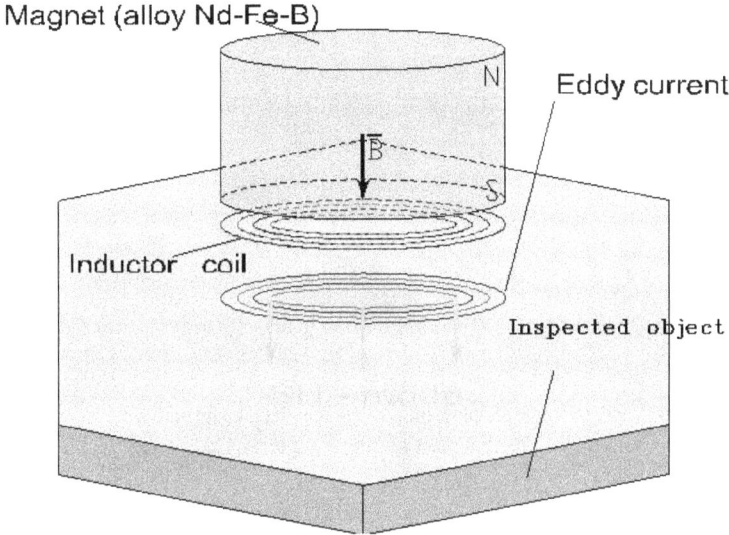

Figure 4.1: EMAT direct-beam technology principle.

Diakont's EMAT technology is capable of inducing the eddy current at multiple angles depending on the goals of the inspection. Direct-beam EMAT coils induce ultrasonic waves into the pipe wall at a 0° (radial) angle to detect pitting corrosion and determine remaining wall thickness. Angle-beam EMAT induces ultrasonic waves at angles inclined to the pipe's ID surface to scan for pitting corrosion below the direct-beam threshold as well as cracks and SCC.

The EMAT transducers rotate circumferentially throughout the inspection to cover the entire pipe wall. Inspection technicians set reporting thresholds for identifying corrosion or cracks to accommodate each customer. The EMAT sensor readings provide measurement threshold accuracy of ± 0.01 in.

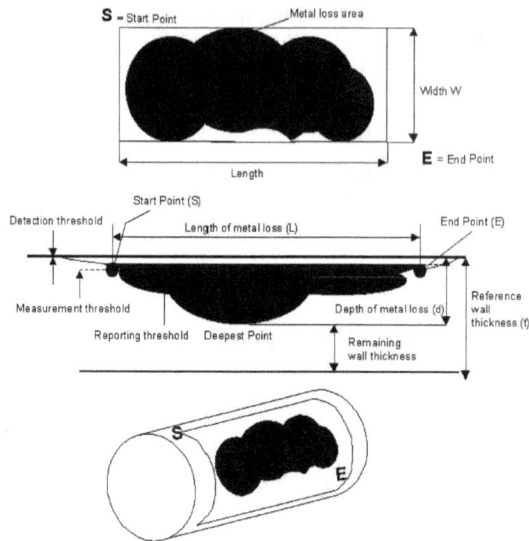

Figure 4.2: An illustration of the general principle of EMAT measurement and reporting thresholds.

Figure 4.3 illustrates an EMAT C-scan, including detection of pipe welds and large pitting corrosion. This methodology supplies structural integrity personnel with crucial information required for planning repairs. All pipe wall assessment data is also stored digitally to provide comprehensive data for monitoring corrosive areas. This baseline assessment information helps pipeline operators manage their assets more effectively through re-inspection and future capital planning.

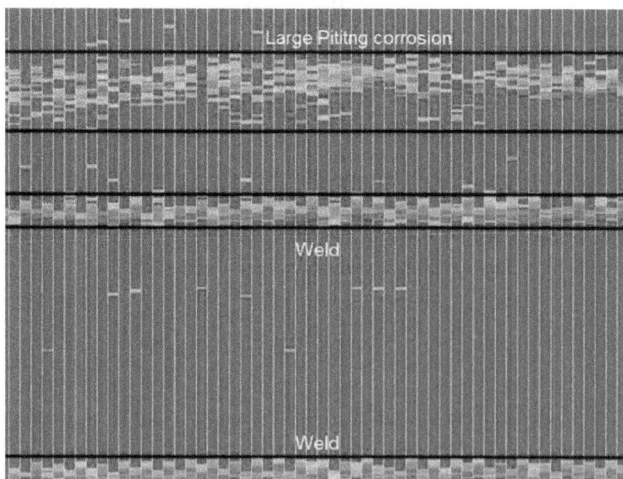

Figure 4.3: Internal pipe anomaly mapping.

# In-Line Inspection and Results

The operator excavated two access points for the inspection to permit crawler tool entry. Diakont mobilized an inspection team to operate the tool and monitor data. The team loaded the crawler into the pipe via removed pipe spools and navigated the inspection area while taking video of the interior and surveying pipe elements. Technicians then returned the crawler to the launch point and began a direct-beam EMAT inspection in automatic scan mode.

After completing the automatic scan, inspection technicians manually drove the robotic crawlers to each indication of a potential defect for detailed characterization. Throughout the inspection, the technicians reviewed the data in real time, monitored tool operation via telemetry, and recorded precise pipeline geometric data, which was incorporated into a pipe tally in the inspection report.

The inspection was completed over a period of five working days—one day less than originally projected. The integrity assessment showed the HCA pipe to be in excellent condition, with only a few minor flaws. Having validated the integrity of the pipeline, the pipeline operator was able to return it to service promptly.

*This chapter is based on and imagery is taken from NACE International Technical Paper C2018-11259, "Robotic Crawler ILI of Unpiggable 10" Natural Gas Pipeline in an Urban Area," by Jonny Minder, Diakont. To view the paper in its entirety, visit store.nace.org.*

# Automation for More Accurate CP Site Surveys

*All cathodic protection site surveys require manual labor, with instrument and/or human errors often skewing results. In fact, data points to almost 90% of survey errors being introduced by humans due a variety of factors including physical stress, lack of proper knowledge or training, improper handling of the survey equipment, etc. A conceptual robot design was developed to conduct CP site surveys in difficult and dangerous on shore and off shore locations.*

Joffin George
SSE Tech Pvt. Ltd.

While performing surveys such as surface inspection for cracks, direct current voltage gradient (DCVG), and close interval potential survey (CIPS), the surveyor has to reach out over the entire metallic structure or the entire pipeline route, taking readings at regular intervals with ultrasonic probes or copper sulfate reference electrodes (CSEs) positioned one in front of the other. This makes the surveys tedious and the surveyor is subjected to various physical and mental challenges, potentially leading to human errors while performing the survey. Human beings have physical limits but machines can be pushed to extreme limits. A conceptual model of a manipulator robot was designed to overcome the above-mentioned challenges, thereby reducing human stress and moreover be able to obtain better survey results within a short duration to time.

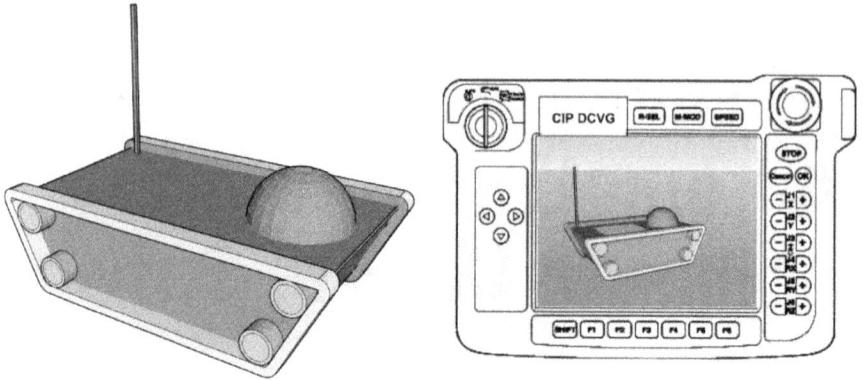

Figure 5.1: Conceptual model of the survey robot
(A) and handheld teach pendant to control the robot from a remote location (B).

The conceptual model was designed by taking into consideration that the robot should perform crack detection, CIP, DCVG, and GPS mapping surveys over on-shore and off-shore structures. The navigation and controlling of the robot can be done from a remote location by the user using the wireless teach pendant.

# Crack Detection on a Flight Deck

An aircraft carrier flight deck is one of the most dangerous work environments in the world. The deck might look like an ordinary runway, but planes are landing and taking off simultaneously at a furious rate in a compact space and one careless act can lead to catastrophic effects. Unlike normal runways, the flight deck isn't long enough to accommodate ordinary landings or takeoffs and hence regular inspection and maintenance is required to ensure they remain free from foreign objects, cracks, holes, and any kind of wear and tear. Even a small nut or bolt, if picked up into an aircraft's engine by suction can cause serious issues.

For example, the 2000 crash of the Concorde in Paris was caused when the aircraft ran over a titanium alloy strip during takeoff, which was left over in the runway by a previous aircraft. The strip blew the aircraft's wheel and punctured the fuel tank, resulting in the death of 113 people. Many airports have patrol teams who look for foreign objects and execute their removal on a day-to-day basis. But the cracks still remain invisible to the human eye.

The above is a particularly well-suited application for the manipulator robot under discussion here. The robot is programmed to automatically traverse the entire run-way looking for any foreign objects and/or cracks. An integrated camera captures images of the runway surface and processes the images to detect the presence of foreign metallic and non-metallic objects and sends live survey information to the handheld teach pendant. The robot also uses ultrasonic probes to detect the cracks in the runway.

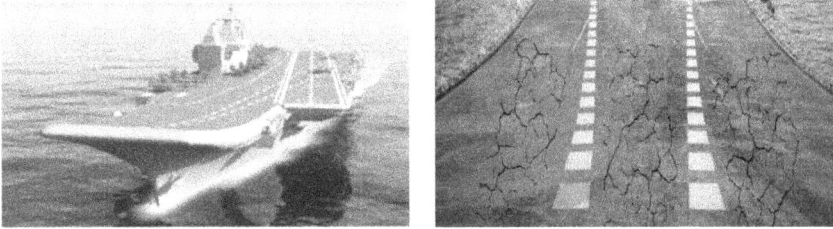

Figure 5.2: Three-dimensional illustration of an aircraft carrier fight deck (A) and cracks found on the runway (B).

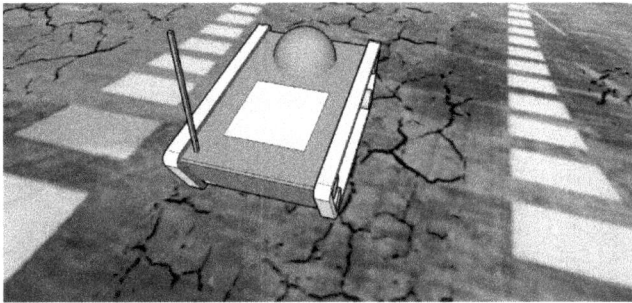

Figure 5.3: Robot performing ultrasonic crack detection on aircraft carrier flight deck.

The ultrasonic crack detector consists of pulser/receiver, ultrasonic transducer, and a display unit. The pulser/receiver produces high voltage pulses and the ultrasonic transducer converts these pulses into high-frequency ultrasonic energy. The sound energy from the transducer is introduced into the test surface, which propagates through the material in the form of waves. When there is a discontinuity or a void in the wave path, part of the energy signal will be reflected back from the flaw surface (Figure 5.4B). The transducer converts the reflected wave signal into electrical signal and is displayed on the screen. The reflected signal strength is displayed versus the time (Figure 5.4A) from signal generation to when an echo was received. The time taken by the signal to travel is directly proportional to the distance that the signal travelled. From the signal, information regarding the reflector location, orientation, size, and other features can be obtained.

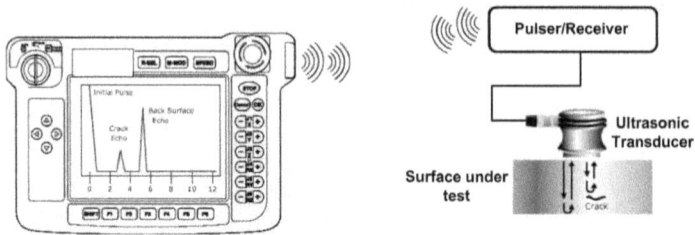

Figure 5.4: Plot of reflected signal versus time in teach pendant
(A) and ultrasonic transducer probe passed over the subject being inspected (B).

# Robot Performing the CIP Survey

During the CIP survey the pipe-to-soil potential (PSP) is recorded at every 1 to 2 m along the entire length of the pipeline by placing the CSE on top of the electrolyte (soil or water). A series of PSPs determines whether adequate cathodic protection is achieved at points along the pipeline route. Figure 5.5 Illustrates the CIPS profile.

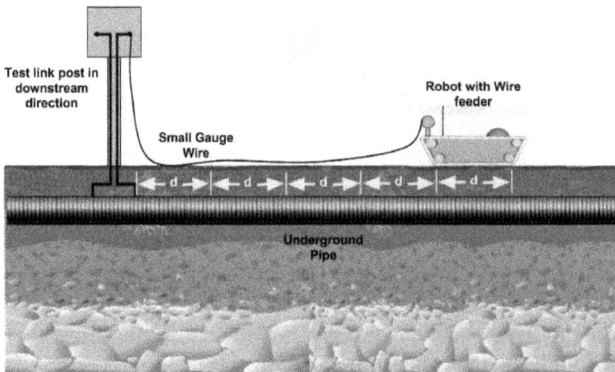

Figure 5.5: Pictorial showing the robot performing a CIP survey.

While performing the CIP survey, pipeline-to-electrolyte potentials are measured and recorded at close equal spacings. When the CIP survey is conducted manually, it is common to find that that the pipeline chain age doesn't synchronize with the CIP survey chain age. This difficulty is overcome using the manipulator robot. The mechanical arrangement in the robot ensures that the PSPs are recorded at uniform close intervals and is integrated with a built-in feedback system to synchronize the actual chain age of the pipeline with the recorded CIP survey chain age. The robot has an inbuilt data logger, a wire dispenser, and a pair of CSEs. The robot is also equipped with an odometer and a GPS unit, so that it knows where it was when the data was entered into the data logger.

Figure 5.6: Odometer pulse wheel under calibration.

Two CSEs are mounted on the bottom portion of the robot (Figure 5.7). A thin coated copper wire of gauge, usually 30 to 40 AWG, is connected to the test lead inside a test link post (TLP) and the user can navigate the robot over the pipeline route, thus making the contact between the electrodes and the ground at regular equally spaced close intervals.

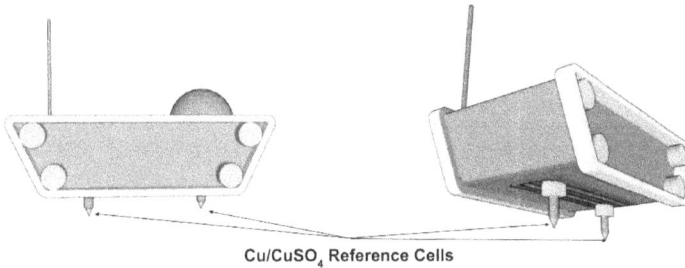

$Cu/CuSO_4$ Reference Cells

Figure 5.7: Side view (A) and bottom view of the robot (B).

The robot stores the GPS, time, and date for each PSP measured during the course of the survey. Live survey data is sent from the robot to the hand-held wireless teach pendant, which enables the user to understand the efficiency of the cathodic protection system applied to the pipeline. The robot uses feedback from various sensors like odometer and GPS units to make sure that duplicate or repeated data is not recorded from a single location.

According to NACE standards, it is recommended to measure the entire PSP by interrupting the cathodic protection applied to the pipeline section and later to repeat the same, to measure the depolarized PSP by turning off the CP system. The intelligent feedback systems in the robot make sure that both the surveys are conducted with the chain ages synchronized. The robot then overlays both the data in the same graph to confirm whether the 100 mV criteria is achieved or not.

# Robot Performing DCVG Survey

The robot used the DCVG technique for detecting the cathodic protection current pick up at coating break downs, often referred to as coating holidays. The robot uses two CSEs along with a digital voltmeter for conducting the DCVG survey over the pipeline. A current interrupter is connected in series in either leg of DC output of the transformer rectifier (TR). The current interrupter creates voltage gradients (Figure 5.8) in the electrolyte which propagates from the coating holidays toward the remote earth. If the pipeline is energized by two or more TR, it is recommended to use GPS to synchronize current interrupters so that all the interrupters trigger at the same time. The interrupters are programmed to trigger at a very fast rate with the ON interval less than the OFF interval. For example, 1/3 seconds ON and 2/3 seconds OFF.

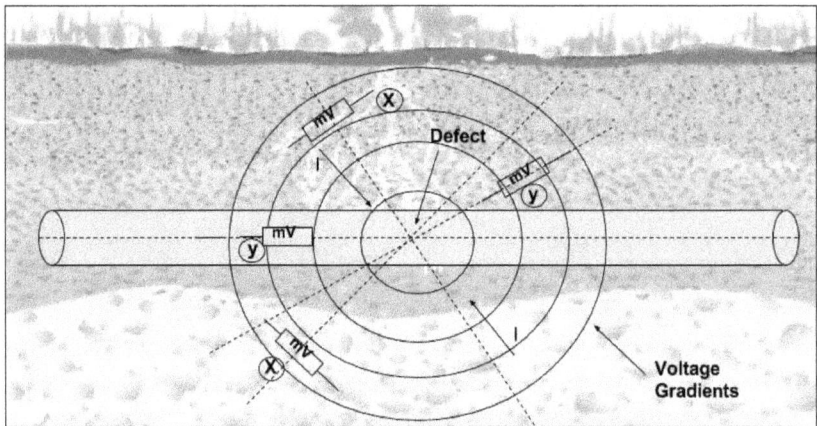

Figure 5.8: Voltage gradients in the soil.

The robot is equipped with a high-input digital voltmeter capable of showing both the positive and negative directions from the zero position of the defect, which helps the robot in understanding the direction of current flowing through the soil. One of the field requirements while carrying out a DCVG survey using the robot is that the ON and OFF potentials from the respective test link posts (TLP) along the pipeline route should be manually measured with respect to CSE with a high input voltmeter and should be stored in the teach pendant. When a defect location is approached the direction of arrow in the screen points toward the defect (Figure 5.9) and when the robot passes the defect location the direction of arrow is reversed.

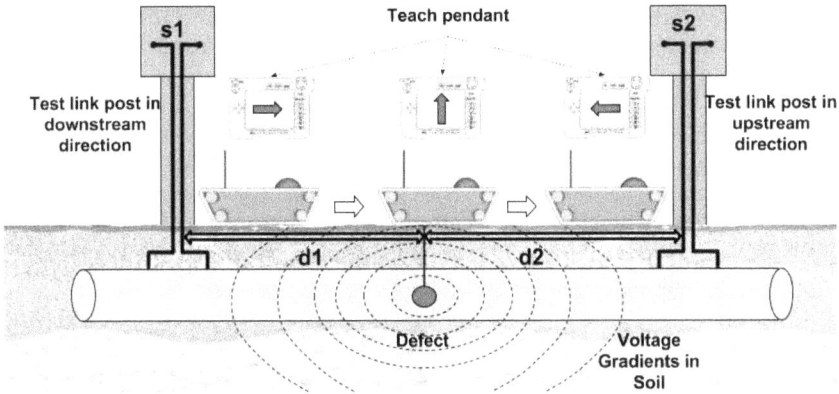

Figure 5.9: Robot performing DCVG survey and the
defect location arrow indication in the teach pendant.

# Computation of Percentage IR by the Robot

The percentage IR is the factor that denotes the size of the coating holiday and is expressed in percentage (%). In order to calculate percentage IR, the robot requires the following parameters.

$d1$  = distance between the defect location and the preceding TLP (meter)
$d2$  = distance between the defect location and the succeeding TLP (meter)
$s1$  = potential shift in the preceding TLP with respect to defect location (mV)
$s2$  = potential shift in the succeeding TLP with respect to defect location (mV) OLRE = over line to remote earth potential (mV)
$S$   = signal strength at the defect (mV)

When the robot starts from chain age zero, the inbuilt odometer gets incremented and when the robot reaches the first TLP the robot stores the corresponding chain age from odometer as d1. The user feeds the potential shift from the first TLP to the robot. The robot continues to move further until it encounters a defect location. The robot calculates dx by subtracting d1 from the current chain age obtained from the odometer. The centering of the defect is done by marking the approximate location of the coating holiday in the area where maximum amplitude is obtained. At two distinct locations, $x$ and $y$ as shown in the Figure 5.8, the reference electrodes are positioned along the voltage gradient until a zero null value is identified. A right angle line through the center of the probe positions shall pass over the coating holiday epicenter. This geometrical procedure, when automatically repeated on opposite sides of the pipeline will give the exact point over the defect. This procedure carried out by the robot is illustrated in the Figure 5.9.

Once the robot identifies the epicenter of the coating defect, the GPS coordinates of the coating defect location is stored and a series of lateral readings are taken by moving towards the remote earth. The robot continues to record the measurements until a one mill volt deflection is obtained, where the robot is assumed to reach remote earth. The summation of these lateral readings are referred to as over line to remote earth (OLRE) potential. Once the OLRE is calculated the robot moves further and identifies another defect and repeats the same procedure until it encounters the second TLP. The current chain age from the odometer is stored as d2. For the upcoming defect locations the preceding d2 becomes d1 and the chain age of the succeeding TLP becomes d2. The signal strength at the defect location (S) is calculated according to Equation 1.

$$S = s1 - \left[\frac{d1}{d1 + d2}\right][s1 - s2] \tag{1}$$

Once the signal strength at the defect location (S) is calculated the robot uses it to calculate percentage IR (Equation 2).

$$\text{Percentage IR} = \left(\frac{\text{OLRE}}{\text{S}}\right) * 100 \tag{2}$$

Theoretically, percentage IR gives the user an idea about the reduction in potential levels without taking into account the polarization effect.

# GPS mapping

During cathodic protection site surveys it is extremely important to precisely locate structures and coating defects along the pipeline route. Since the pipeline can extend from a few to many kilometers, the only way it can be realized is by using a GPS. The robot is integrated with a highly accurate GPS receiver antenna as shown in Figure 5.10B, because precise position of the defect location and pipeline route is particularly important such that the cost of labor and machine can be drastically reduced.

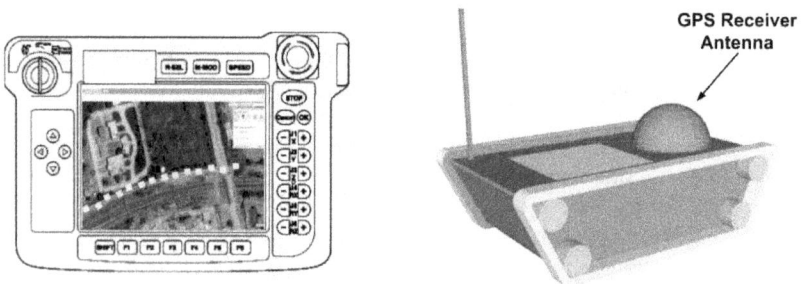

Figure 5.10: A) GPS mapping displayed in the teach pendant and B) position of the GPS receiver antenna on the robot.

Figure 5.11: With the design phase for the manipulator robot for cathodic protection surveys complete, shown is construction of the hardware model underway. Future work will include real time testing of the robot in the site and its corresponding results.

To carry out such tasks the manipulator robot will perform to be the best economic solution. The accuracy of the inbuilt GPS receiver unit is less than one meter. The device is programmed in such a manner that it continuously receives a correction signal from satellites, thus the sub-meter precision is provided at the survey site and no further processing is required. The user is able to track the robot using the teach pendant from remote location (Figure 5.10A). All these operations contribute to ensure safe and healthy working environment, meanwhile the operating cost is significantly reduced.

*This chapter is based on and imagery is taken from NACE International Technical Paper C2018-11210, "Manipulator Robot for Cathodic Protection Site Survey," by Joffin George, SSE Tech Pvt. Ltd. To view the paper in its entirety, visit store.nace.org.*

# UAV's Take on Mission for Marine Corrosion, Coating Inspections

*New partnerships aim to expand drone use for maritime asset maintenance. The concept of using UAVs for inspections is of particular interest to the maritime industry, since the marine environment presents numerous spaces that require either significant human risk or significant financial cost to access. However, as with many new technologies, challenges come with commercialization, costs, and processes.*

Ben DuBose
Materials Performance

Corrosion has long been a major challenge for marine asset owners and operators, with corrosion under insulation (CUI) near the forefront of the list. The abundant moisture found near marine assets can often become trapped around pieces of equipment and even within insulated material—leading to accelerated localized corrosion of the underlying metal substrate. Galvanic corrosion is widely considered the primary form of corrosion here.

In recent years, the idea of using unmanned aerial vehicles (UAVs)—or drones—for surveillance operations to help spot such problems has surged in popularity across many industries, led by oil and gas and others prone to the effects of corrosion. Traditionally, these inspections have been carried out by human crew members, surveyors, or independent inspectors—a risky activity that represents one of the most common causes of work-related industry fatalities. That risk is often further heightened in marine and offshore environments.

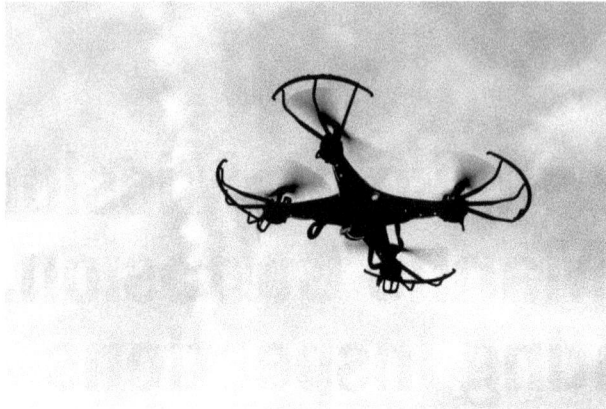

Figure 6.1: Various devices, such as high-definition video and cameras, are attached to drones and used to detect corrosion.

Besides the safety issue, these traditional practices may not always be completely effective. In the example of CUI, since removing all insulation material and examining the substrate underneath is cost prohibitive, the usual practice is to remove small portions of the insulation at select locations that could be at risk, and then use nondestructive testing (NDT) techniques on the surface to determine if there is metal loss. That practice, however, can sometimes spark a new problem by creating a potential entry point for moisture ingress. In addition, since CUI is localized, corrosion may not be found if the point where the insulation is removed is not covering the specific area where corrosion is occurring.

UAVs, however, have shown the potential to help the process in multiple ways. First, they can access difficult, hard-to-access environments, which reduces the safety risk for human inspectors. Second, by using remote thermal infrared (IR) and multispectral imaging sensors, they can detect anomalies that can be indicative of corrosion—even without removing the insulation or the existing coating.

The concept of using UAVs for inspections is of particular interest to the maritime industry, since the marine environment presents numerous spaces that require either significant human risk or significant financial cost to access. However, as with many new technologies, challenges come with commercialization, costs, and processes.

To address those questions, a number of research and development (R&D) projects launched in recent months are aimed at facilitating a more widespread adoption of drone (UAV) use to inspect for problems such as corrosion. These projects involve partnerships between industry, academia, and drone technology groups—all designed to develop new end-to-end processes to enable the frequent use of drones to perform inspections in maritime settings.

# Enclosed Spaces, Ballast Water Tanks

One such collaboration announced earlier this year comprises global paints and coatings company AkzoNobel, oil and gas tanker operator Barrier Group, and DroneOps. Given the code name RECOMMS (Remote Evaluation of Coatings and Corrosion on Offshore Marine Structures and Ships), their project aims to use the semiautonomous operation of a drone to assist with coating and corrosion checks.

With a goal of boosting safety, project officials say their drone will use advanced virtual reality technology to deliver safer, more accurate evaluations of many enclosed or difficult-to-access spaces on ships and marine structures, including ballast water tanks. Those evaluations will involve inspections of corrosion damage and any deterioration of the existing coating. The drone should be able to detect CUI and help identify the need for maintenance further in advance, according to the developers.

Figure 6.2: AkzoNobel, Barrier Group, and DroneOps joined a wave of collaborations aimed at promoting drone use to help with marine maintenance issues. Photo courtesy of AkzoNobel.

"Surveys of enclosed spaces and ballast water tanks are an essential part of routine maintenance and are increasingly critical for ship owners," said Michael Hindmarsh, business development manager for AkzoNobel's International Paint marine coatings business. "Inspecting these areas thoroughly can require working at height, entering confined spaces, and negotiating slippery surfaces that could be poorly lit, all of which are high-risk activities that the maritime industry is keen to address."

By replacing human inspections with a drone, routine maintenance can be monitored remotely and in real time by office-based staff, with instant feedback available to the vessel or offshore structure's superintendent, according to the companies. In turn, this can reduce costs, increase efficiency, and significantly reduce risks to human workers during essential maintenance.

The companies explain that the partnership's experience, which includes coatings expertise and consultancy, drone building, ownership of marine structures, and a working knowledge of present repair and inspection practices, provides a complete overview of issues and challenges associated with enclosed space inspections. As part of the partnership, additional coatings information will be provided by coatings consultancy Safinah Ltd.

AkzoNobel notes that is already using drone technology. The company is currently testing the use of drones in Australia for inspecting sites in remote locations—where access is limited and the movement of heavy equipment is difficult. Thus far, the results have shown significant promise, according to Hindmarsh, with specific findings expected to be published later this year.

As the consortium pushes forward, the drone will undergo flight trials at an existing coatings test site in the U.K. and also at an indoor training facility run by the tanker operator.

## 'Caged' Drone

The RECOMMS project is the latest in the string of marine drone announcements. Some time ago, in November 2016, Robotica in Maintenance Strategies (RIMS) officially launched its own new service for the maritime and offshore industries using a drone with a protective cage, named "Elios."

"We carried out extensive market research including visiting several universities in Holland and Switzerland with our partner Flyability, where they gave a presentation of their drone Elios," said Senior Maintenance Engineer David Knuckell. "This is a drone within a protective cage, and is perfectly suited to enter enclosed spaces and carry out in-depth inspections of the enclosed areas."

Figure 6.3: One marine consortium led by Robotica in Maintenance Strategies (RIMS) has proposed enclosing their drone within a protective cage. Photo courtesy of Flyability.

The cage enables Elios to "bounce of walls" and "fly where no other drone can," according to Flyability, which used the drone in a case study last fall to inspect a storage tank for oil and gas terminal operator Royal Vopak.

"We used to climb down, had to arrange all kinds of safety measures for people, and had to light up lamps to do the inspection," said Jan Zandberg, a terminal manager at Vopak. "Now we can use the drone, which saves us an enormous amount of time, but also lowers the risk of sending people down there. We now inspect the whole tank in about two hours. In earlier days, people had to go down, all kinds of precautions had to be made. This cost days."

She added that "this technology has just started. We see huge advantages to using this technique in the future. It will become a mainstream technology, I'm absolutely convinced of that."

For projects involving the Elios, the U.K.-based drone developer is working with drone-based inspection services provider Sky-Futures to bundle the drone with Expanse software for interpretation of the findings. The software helps make data available to all stakeholders through a cloud system, and it enables clients to present inspection findings in a three-dimensional (3D) environment. According to the companies, the combination of the drone and software allows for an end-to-end solution of data capture, processing, and distribution.

"Through this bundle package, we intend to provide our customers with the greatest flexibility and efficiency in the way they can disseminate and post-process data gathered," said Patrick Thévoz, CEO and co-founder of Elios.

In June 2016, the two companies completed the world's first trial inspection of a floating production, storage, and offloading (FPSO) cargo tank by drone without using a human to enter the tank. According to Karen Cowie, BW Offshore's senior integrity engineer for the Athena FPSO in the U.K.'s North Sea, the drone was able to fly down into the tank unaided and accurately navigate the internal space for inspections.

"From the inspection completed, it is clear that the benefits in terms of not just time and cost to inspect but also preparation, cleaning, repeatability, and access requirements highlight that this technology is an exceptional tool to have available," said Cowie. "For our specific requirements, the safety benefit to be gained by avoiding personnel entry is invaluable."

## Preprogrammed Inspection Missions

In September 2016, technical services and maritime classification organization Lloyd's Register signed an agreement with drone company Airobotics to help develop its own use of remote-access technologies, including the use of autonomous drones for preprogrammed inspection missions.

In this collaboration, the drone company has developed a fully automatic platform that is continuously available on-site and enables both preprogrammed missions and expert access in an on-demand format. The platform is self-sustained, with an ability to replace its own batteries and payloads as required for different missions.

According to project officials, the payloads used on drone inspections include various capture devices such as high-definition video cameras that capture still shots for IR thermography, which is used frequently to detect CUI. Further technical capabilities available with their platform include 3D models and maps generation, with other sensors available upon request.

Figure 6.4: Under one proposal, data collected by drones could potentially enable engineers to monitor marine structures at a remote, land-based control center. Photo courtesy of Rolls-Royce.

Looking ahead, the groups plan to focus on additional development within the platform's hyperspectral capabilities. Hyperspectral imaging works by obtaining the full spectrum for each pixel in an image, with a goal of covering a wider range of wavelengths than can be seen by the human eye and better detecting materials or patterns in the object being inspected.

"We believe our cooperation will open doors for the maritime industry to reveal a new level of efficiency and innovation with our automated, industrial-grade drone solution," said Ran Krauss, CEO of Airobotics.

# Land-Based Control Center

In March 2016 integrated power and propulsion solutions provider Rolls-Royce unveiled its vision of a land-based control center to remotely monitor and control unmanned ships. In a six-minute demonstration video, the company showed how the program would use surveillance drones to help monitor what is happening around a ship.

"We're living in an ever-changing world, where unmanned and remote-controlled transportation systems will become a common feature of human life," said Iiro Lindborg, general manager of the company's remote and autonomous operations segment within its ship intelligence division. "They offer unprecedented flexibility and operational efficiency."

He added that his company's "research aims to understand the human factors involved in monitoring and operating ships remotely. It identifies ways crews ashore can use tools to get a realistic feel for what is happening at sea."

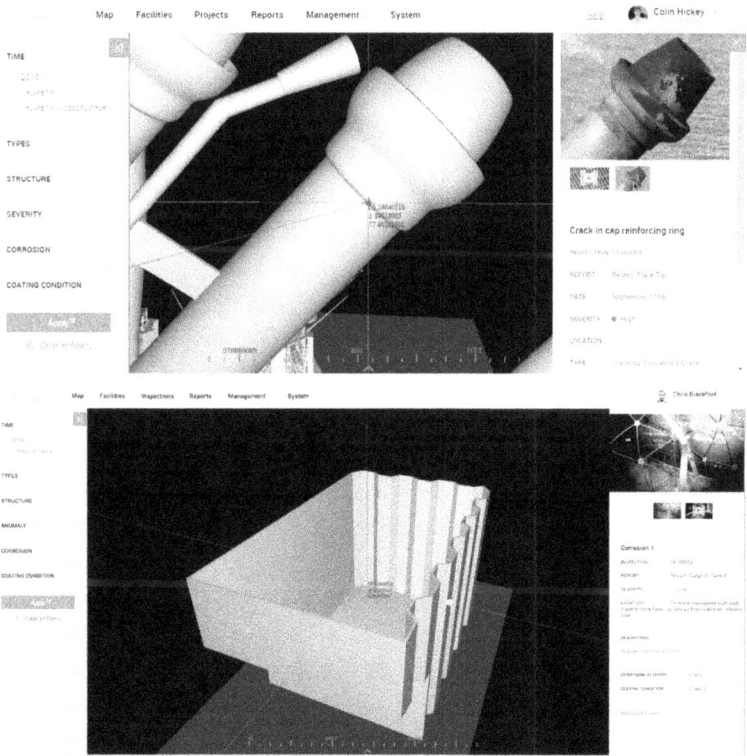

Figure 6.5: Drone firm Flyability is working with an inspection services provider to bundle drones with Expanse software to improve interpretation of the findings. Photo courtesy of Sky-Futures.

Partners on the land-based control center project include nonprofit R&D group VTT and the Tampere Unit for Computer Human Interaction (TAUCHI) research unit at the University of Tampere.

Their project envisions using staff at a control center to plot a complete course for autonomous vessels—each with remotely piloted drones for inspection and predictive maintenance operations—before turning the process over to regional remote operators. The technology would enable a small crew of between seven and 14 people to monitor and control an entire fleet.

The groups involved in this joint research project plan to build a project demonstrator "before the end of the decade."

*This chapter is based on the* Materials Performance *article, "Drone Shows Promise in Measuring Coating Thickness," by Ben Dubose. To read the article in its entirety, visit www.materialsperformance.com.*